Smart Waste Management

Harnessing Technology for a Sustainable Tomorrow

Leopold Morton

1

Smart Waste Management

TABLE OF CONTENTS

Chapter 1: Understanding Smart Waste Management

The Evolution of Waste Management Practices

Waste management has undergone a significant transformation, evolving from rudimentary practices to sophisticated systems designed to address the complexities of modern urban living. This evolution reflects not only technological advancements but also shifts in environmental awareness, policy changes, and societal expectations. Understanding the journey of waste management allows us to appreciate the systems in place today and the importance of continual improvement for future sustainability.

In ancient civilizations, waste disposal was a straightforward affair. Early societies managed waste through basic means, often relying on natural processes. Waste, such as organic material, was left to decompose in the environment, while inorganic items were typically discarded in pits or near dwellings. The Romans are often credited with creating the first organized waste management systems, establishing regulations for garbage collection and the development of public latrines. These early efforts marked the recognition of waste disposal's importance to public health.

With the advent of the Industrial Revolution in the 18th and 19th centuries, urbanization took off, leading to an explosion in waste generation. Cities grew rapidly, and people moved to urban centers for work, creating a pressing need for more structured waste management systems. Unfortunately, this growth often outpaced the infrastructure needed to deal with the resulting waste. Streets became clogged with garbage, leading to unsanitary conditions and public health crises. In this environment, municipalities began formalizing waste collection and disposal practices, paving the way for the modern waste management industry.

The rise of landfilling in the 20th century marked a significant turning point. Initially seen as a simple solution, landfills quickly presented challenges, including odor, vermin, and groundwater contamination. These issues highlighted the need for more sustainable practices. Recognizing the environmental impact of landfills, the focus began shifting toward reducing waste through recycling and composting initiatives.

The mid-20th century brought further technological advancements that significantly changed waste management practices. The introduction of mechanical systems for collection and sorting enabled more efficient waste processing. This period saw the rise of recycling programs, spurred by environmental movements and increased public awareness of pollution. Various communities implemented source separation, encouraging residents to categorize their waste into recyclables, organic material, and trash. This concept of source separation was revolutionary, as it connected consumers directly to the environmental consequences of their consumption habits.

During the 1990s and early 2000s, the global narrative surrounding waste management began emphasizing sustainability and circular economy principles. The traditional linear model of "take, make, dispose" was increasingly scrutinized, giving way to a more circular approach that seeks to minimize waste through reuse, recycling, and recovery. This transition also prompted policymakers to enact stricter regulations governing waste generation and disposal. The European Union's Waste Framework Directive, for instance, aimed to reduce waste and promote recycling, setting ambitious targets for member states.

As technology advanced into the 21st century, innovative strategies began reshaping the landscape of waste management. The development of smart technologies, such as IoT devices, enabled real-time monitoring of waste collection and processing. Cities adopted smart bins equipped with sensors that detect fill levels, optimizing collection routes and reducing operational costs. Such innovations allow municipalities to allocate resources more effectively, resulting in cleaner streets and enhanced service delivery.

The integration of data analytics also revolutionized waste management practices. By analyzing waste generation patterns and materials recovery data, cities have gained valuable insights that inform better decision-making. Predictive analytics helps forecast waste generation during events or peak seasons, allowing for proactive measures and increased efficiency. This data-driven approach ensures that waste management systems are not only reactive but also strategically planned for future growth and resilience.

Education and community engagement play crucial roles in the evolving landscape of waste management. Public awareness campaigns aimed at promoting recycling and composting have empowered individuals to make informed choices about waste disposal. Educational programs in schools encourage the next generation to adopt sustainable habits early, fostering a culture of environmental responsibility. By involving citizens in waste management processes, communities can take ownership and contribute to achieving local sustainability goals.

However, implementing these advanced practices is not without challenges. One of the primary obstacles is the need for significant investment in infrastructure and technology. Many cities lack the financial resources to upgrade aging waste management systems, which can hinder the adoption of innovative solutions. Additionally, there is often resistance to change from various stakeholders, including waste management companies and residents accustomed to traditional practices.

To address these challenges, collaboration across different sectors is essential. Partnerships between governments, private enterprises, and non-profit organizations can create synergies that drive innovation and investment in waste management infrastructure. Successful case studies abound where collaborative approaches have led to improved waste management outcomes, demonstrating the potential for collective impact when diverse stakeholders work together.

Moreover, regulations and policies must evolve in tandem with technological advancements. Policymakers are tasked with creating a supportive legislative environment that encourages innovation while safeguarding public health and the environment. This includes establishing clear guidelines for waste reduction targets, recycling rates, and the use of emerging technologies in waste management processes.

The future of waste management must embrace a holistic approach that considers economic, environmental, and social factors. As urban populations continue to rise, the pressure on waste management systems will intensify. Emphasizing the importance of waste minimization, resource recovery, and community engagement will be vital in crafting resilient waste management strategies. Engaging citizens in the process cultivates a sense of responsibility and encourages mindful consumption habits, which are integral to reducing overall waste generation.

Ultimately, the evolution of waste management practices reflects society's shifting values regarding waste. The early disregard for waste's environmental impact has transformed into a recognition of our collective responsibility to manage waste sustainably. Each step taken, whether through technological innovation or community engagement, contributes to a larger narrative of sustainability and resilience.

As we stand at this crossroads, it becomes clear that the ongoing evolution of waste management is not merely a response to challenges but also an opportunity to rethink how we interact with our environment. By embracing innovation, fostering collaboration, and committing to sustainable practices, we can forge a path toward a cleaner, healthier future for our communities and the planet.

Defining "Smart" in Waste Management Contexts

The term "smart" has permeated various industries, reflecting a world increasingly driven by technology and data. In the context of waste management, defining what constitutes "smart" goes beyond simply integrating new gadgets or software. It embodies a fundamental shift towards efficiency, sustainability, and community engagement. Smart waste management encompasses a suite of innovative practices that leverage technology to improve waste collection, processing, and disposal methods, ultimately aiming for a more sustainable future.

At its core, smart waste management leverages the Internet of Things (IoT). This network of interconnected devices provides real-time data, enabling municipalities to monitor waste levels in containers, track collection routes, and optimize scheduling. Imagine a city where waste bins equipped with sensors signal when they are full, allowing waste collection teams to plan their routes accordingly. This proactive approach not only minimizes operational costs but also reduces unnecessary carbon emissions from collection vehicles, thereby promoting environmental sustainability.

Central to smart waste management is the integration of data analytics. By harnessing big data, cities can identify patterns and trends in waste generation. Understanding when and where waste is produced most allows for more strategic resource allocation. For example, during festival seasons or large community events, waste generation spikes. With predictive analytics, municipalities can anticipate these surges and adjust their services accordingly, ensuring that waste is managed effectively while enhancing public satisfaction.

Community engagement plays a crucial role in smart waste practices. Technology must complement the active participation of residents in waste reduction and recycling efforts. Educational initiatives can inform citizens about proper disposal methods, recycling guidelines, and the importance of minimizing waste. Digital platforms, including mobile applications, can facilitate this engagement by providing users with information on local recycling programs, drop-off locations for hazardous waste, and real-time alerts about collection schedules. When communities are empowered with knowledge, they are more likely to adopt sustainable habits, creating a culture of responsibility toward waste management.

The benefits of adopting smart technologies in waste management extend far beyond operational efficiencies. Economically, cities that invest in smart systems can realize substantial cost savings. By optimizing collection routes and reducing fuel consumption, municipalities can allocate funds to other pressing needs, such as public services or infrastructure improvements. Additionally, the increase in recycling rates can lead to revenue generation from recovered materials, creating a financial incentive for cities to go green.

However, the implementation of smart waste management systems is not without challenges. The initial investment in technology and infrastructure can be significant, posing a barrier for many municipalities, especially those with limited budgets. Furthermore, the successful deployment of smart technologies necessitates reliable internet connectivity, which may not be available in all areas. Addressing these issues requires a collaborative approach among governments, private enterprises, and community stakeholders to create a supportive ecosystem for smart waste management.

Moreover, the definition of "smart" must also encompass the adaptability of waste management systems. As urban populations grow and consumption patterns shift, waste management strategies must evolve. Smart technologies should be user-friendly and scalable to accommodate these changes. For instance, cities can pilot smart waste initiatives in specific neighborhoods before expanding them citywide. This phased approach allows for the refinement of systems based on real-world feedback and results, ultimately leading to more effective and widely accepted solutions.

Public policy frameworks must support the transition to smart waste management. Policymakers have a critical role in establishing regulations that promote innovation while ensuring public health and environmental safety. Incentives for adopting smart practices can encourage municipalities to invest in technology. Additionally, regulations that align with sustainable development goals can drive collective action toward waste reduction and resource recovery.

As the notion of "smart" evolves, it becomes essential to measure its impact. Metrics such as recycling rates, waste diversion rates, and community engagement levels can help gauge the effectiveness of smart waste management initiatives. Regular reporting and transparency in these metrics not only keep stakeholders informed but also foster public trust. When communities see the tangible benefits of smart practices, they are more likely to participate actively and support ongoing efforts.

One effective model for implementing smart waste management is the concept of circular economy principles. This model emphasizes the continuous use of resources, aiming to keep materials in circulation for as long as possible. Through smart technologies, cities can facilitate efficient material recovery, ensuring that valuable resources are not lost to landfills. For example, integrating waste sorting robots in recycling facilities can enhance the quality and quantity of recyclable materials, driving the circular model further.

Another critical aspect of smart waste management is its potential to address social equity concerns. Ensuring that all communities have access to smart waste services is essential for creating inclusive and effective systems. This may involve deploying technology in underserved areas, ensuring that education and resources are accessible to all citizens, regardless of socioeconomic status. When smart waste management is equitable, it fosters community involvement and resilience, allowing every member of a society to contribute to sustainable practices.

The future of smart waste management lies in continuous innovation and collaboration. Emerging technologies such as artificial intelligence and machine learning hold vast potential for streamlining waste processing and enhancing decision-making. As these tools become more sophisticated, they can offer predictive insights and automate processes that once required extensive human labor. Implementing such technologies can further improve operational efficiency and environmental outcomes.

Ultimately, defining "smart" in waste management contexts involves a multifaceted approach. It requires the integration of advanced technologies, meaningful community engagement, and supportive policy frameworks aimed at sustainability and efficiency. By embracing these principles, we can transform our waste management practices, moving toward a more sustainable and resilient future. As cities navigate the complexities of waste generation in an increasingly urbanized world, the adoption of smart practices will be crucial in managing this challenge effectively.

We find ourselves at a pivotal moment where the collective actions of municipalities, community members, and businesses can lead to significant progress. Embracing the concept of smart waste management is not merely a trend; it is an essential step toward ensuring that we leave a healthier planet for future generations. In this endeavor, every small action, whether through education, technology adoption, or community involvement, contributes to a broader movement toward a responsible and sustainable approach to waste. Each effort reinforces the foundation of smart waste management, paving the way for innovations that can transform our relationship with waste and materials in a rapidly changing world.

The Role of Technology in Modern Waste Solutions

The landscape of waste management is shifting rapidly, driven by the need for more efficient, sustainable, and innovative solutions. Technology serves as a catalyst for this change, introducing methodologies that not only enhance efficiency but also ensure that waste management aligns with environmental stewardship. As the world grapples with increasing waste generation, the integration of modern technology in waste solutions is more crucial than ever.

Advanced data analytics is paving the way for proactive waste management strategies. Cities are utilizing historical data to discern patterns in waste production, allowing for the anticipation of peak waste periods. For instance, if a city notices elevated waste levels during holiday seasons, it can adjust collection frequencies and resources in advance. This foresight minimizes overflow situations and ensures efficient operations, reducing the need for emergency responses that can strain resources and finances.

Smart bins equipped with sensors are another revolutionary development. These bins can monitor fill levels and transmit real-time data to waste management operators. By knowing when containers are full, collection routes can be optimized, reducing fuel consumption and diminishing vehicle emissions. This technology not only enhances operational efficiency but also significantly lowers the carbon footprint associated with waste collection. Moreover, when urban planners use these insights, they can strategically place bins in high-traffic areas, ensuring access and encouraging recycling.

Automated sorting technologies further exemplify how modern solutions can transform waste management. Traditional facilities often rely heavily on manual sorting, which is labor-intensive and prone to human error. The introduction of robotic systems equipped with machine vision allows for high-speed, precise sorting of recyclables from general waste. These systems can identify various materials, such as plastics, metals, and paper, ensuring that recyclable materials are diverted from landfills. The result is an increase in recovery rates while simultaneously reducing the operational costs of waste processing.

Additionally, the concept of artificial intelligence in waste management is gaining traction. AI algorithms can analyze historic data on waste generation and recovery, allowing for smarter decision-making in resource allocation and process optimization. For instance, predictive models can estimate future waste flows based on demographic trends or economic indicators, enabling municipalities to scale their operations proactively rather than reactively. The insights gleaned from AI applications facilitate a shift toward a more holistic understanding of waste as a resource rather than mere refuse.

Digital platforms play an equally important role in engaging the community and increasing transparency in waste management practices. Mobile applications allow residents to track waste collection schedules, access recycling guidelines, and report missed pickups. These platforms empower citizens to take an active role in waste management by increasing awareness and fostering community pride. When residents see their contributions—be it through recycling or proper disposal—they are more inclined to participate actively, leading to better overall waste management outcomes.

Educational initiatives supported by technology can enhance community engagement even further. Virtual reality (VR) and augmented reality (AR) experiences can simulate the impact of waste on the environment, providing immersive educational tools that resonate with a broader audience, particularly younger generations. By experiencing the consequences of global waste issues firsthand, individuals may be more motivated to adopt sustainable practices, thus fostering a culture of responsibility.

Challenges still exist, however, when implementing technology in waste management systems. The initial investment for advanced technologies can be substantial, and for many municipalities, this poses a significant hurdle. Securing funding often requires presenting a compelling case for the long-term benefits of these investments. Moreover, as the reliance on technology increases, so does the need for skilled personnel who can operate and maintain these systems. Training and workforce development become critical factors for success.

Collaboration between public and private sectors is essential in overcoming these challenges. By forging partnerships, cities can access the expertise and resources necessary to implement advanced waste technologies effectively. For example, private companies specializing in waste technology can partner with municipal governments to develop tailored solutions that meet specific community needs. These collaborations can also facilitate knowledge sharing, which is crucial in the fast-evolving field of waste management.

The advancement of waste-to-energy technologies is another vital component of modern waste solutions. Converting waste into energy provides a dual benefit: reducing landfill volumes while generating renewable energy. Technological innovations in anaerobic digestion and incineration with energy recovery systems are becoming more prevalent. These systems not only mitigate the environmental hazards associated with landfills but also contribute to energy self-sufficiency, allowing communities to harness waste as a viable resource.

Sustainability is enhanced further through the use of blockchain technology in waste management. This technology can provide transparent and traceable records of recycling activities from inception to completion. By tracking waste materials through each stage of the process, stakeholders can ensure compliance with recycling regulations and bolster accountability. This transparency helps build trust within communities, as residents can see tangible outcomes from their recycling efforts.

Additionally, advancements in biotechnologies are paving the way for new methods of waste treatment. Microbial processes that break down organic waste into compost or biogas create valuable resources while reducing the amount of waste sent to landfills. These innovations not only further sustainability goals but also contribute to the emergence of a circular economy, where materials are continuously reused and repurposed rather than discarded.

As communities adopt technology in waste management, a cultural shift begins to take shape. When technology facilitates efficient operations and engages residents, it cultivates a sense of shared responsibility toward waste reduction. This cultural engagement is essential in the long term; the success of waste management initiatives is deeply rooted in public participation and awareness.

The global landscape is dynamic and varied, with cities around the world facing unique waste management challenges. Learning from successful models is paramount. For instance, cities like San Francisco have established ambitious zero-waste goals, utilizing a combination of technology, policy, and community engagement to drive substantial changes in waste management practices. By studying such models, other municipalities can adapt strategies that resonate with their local context, fostering a more sustainable future.

In conclusion, technology's role in modern waste management cannot be overstated. From data analytics to community engagement platforms, each innovation harbors the potential to transform how we think about waste. The journey toward efficient waste solutions requires collaboration, investment, and a commitment to public stewardship. As we embrace these technological advancements, we move closer to a more sustainable, responsible way of managing our resources, ultimately contributing to healthier communities and a cleaner planet. Embracing technology is not just a necessity; it is an opportunity to redefine our relationship with waste and drive meaningful change.

Environmental and Economic Benefits

The intersection of environmental sustainability and economic growth is a delicate balance that many communities strive to achieve. In today's world, understanding how these two aspects interact can lead to innovative solutions that provide long-lasting benefits. When a community makes concerted efforts to embrace sustainable practices, the advantages extend beyond mere compliance with regulations; they significantly improve the quality of life, enhance local economies, and promote environmental health.

One of the most immediate benefits of adopting sustainable practices is the reduction of waste. By implementing comprehensive recycling and composting programs, communities can divert significant amounts of waste from landfills. This not only conserves valuable landfill space but also decreases the environmental impact associated with waste decomposition, which produces harmful greenhouse gases like methane. Additionally, many cities have seen a substantial reduction in waste management costs when recycling rates improve. The initial investments in recycling infrastructure can be offset by lower disposal costs, demonstrating a direct economic benefit.

Transitioning to renewable energy sources represents another powerful strategy with both environmental and economic rewards. Solar, wind, and other renewables have become more accessible and affordable, allowing communities to reduce their dependence on fossil fuels. The shift to renewable energy not only decreases greenhouse gas emissions but also fosters local job creation. The renewable energy sector has been one of the fastest-growing job markets, offering opportunities in installation, maintenance, and manufacturing. Communities that invest in renewables stand to gain economically while simultaneously addressing pressing environmental concerns.

Local food systems bring additional benefits, promoting environmental sustainability while bolstering the economy. By supporting local farmers through farmers' markets and community-supported agriculture (CSA) programs, communities can reduce the carbon footprint associated with transporting food over long distances. Lower transportation emissions translate into cleaner air and a healthier environment. Moreover, purchasing locally grown food keeps money circulating within the community, enhancing local economies and strengthening community ties. When residents prioritize local produce, they not only gain fresher, healthier options but also contribute to agricultural resilience.

Investing in green infrastructure—such as urban green spaces, green roofs, and permeable pavement—offers multifaceted advantages. These approaches can minimize stormwater runoff, reducing the burden on drainage systems and improving water quality. Clean waterways contribute to healthier ecosystems, which can alleviate the costs associated with water treatment and flood control. Furthermore, green spaces enhance property values and attract tourism, contributing to local economic growth. Communities that nurture their natural landscapes can enjoy improved public health outcomes, as green areas promote physical activity and mental well-being.

Engaging in sustainable transportation initiatives also yields significant economic and environmental benefits. Promoting public transit, cycling, and walking not only reduces reliance on single-occupancy vehicles but also decreases traffic congestion and air pollution. Cities that invest in bike lanes and pedestrian pathways create healthier, more attractive communities. Reduced congestion saves residents time and money while enhancing air quality, leading to improved public health. A vibrant public transit system strengthens local economies by connecting residents to job opportunities and services, fostering accessibility for all demographics.

Implementing energy-efficient technologies can further amplify environmental benefits and lead to substantial cost savings. Retrofitting buildings with energy-efficient systems, such as better insulation, LED lighting, and smart thermostats, reduces energy consumption. Lower energy bills benefit households and businesses, freeing up resources for other expenditures. Energy efficiency initiatives can promote local employment through retrofitting projects, all while decreasing the carbon footprint of the community.

When communities shift toward sustainable practices, they often discover the power of education and community involvement. Initiatives that raise awareness about the importance of sustainability can motivate residents to adopt eco-friendly habits. Educational programs in schools and community centers can teach individuals about waste reduction, energy conservation, and the local food system. Engaged communities are more likely to participate in sustainability efforts, creating a culture that values responsible choices and nurtures a shared commitment to protecting the environment.

Sustainability can also enhance a community's resilience against climate change and unforeseen economic shocks. By focusing on local resources and reducing dependence on external supplies, communities can better withstand fluctuations in global markets. Diverse food systems, renewable energy sources, and localized economies can provide stability and security. Resilience is not just about surviving challenges; it's also about thriving amid them. Investing in sustainability fosters a robust community ready to adapt and innovate in the face of adversity.

The importance of collaboration cannot be overlooked when discussing the benefits of sustainability. Partnerships among public agencies, private businesses, and nonprofit organizations can amplify efforts aimed at reducing waste and promoting economic growth. Collaborations can facilitate the sharing of resources, knowledge, and best practices that fuel innovative solutions. For example, public-private initiatives may focus on developing clean technologies or creating incentives for businesses to adopt greener practices. The collective impact of these partnerships can leave a lasting imprint on both the environment and the local economy.

Moreover, the advancement of technology enhances the efficacy of sustainability initiatives. Digital platforms can streamline recycling processes, facilitate carpooling, and provide residents with data on their energy consumption. Communities can track progress and evaluate the effectiveness of their sustainability strategies. By employing data-driven decision-making, leaders can allocate resources more effectively, prioritize initiatives that yield significant returns, and respond to emerging challenges with agility.

The ongoing transition towards a circular economy epitomizes an innovative approach that provides significant benefits for both the environment and the economy. A circular economy focuses on minimizing waste through the continual repurposing of materials. By designing products with longevity and recyclability in mind, manufacturers can reduce resource extraction and waste generation. This shift can generate new business opportunities and drive innovation, contributing to job creation while mitigating environmental impacts.

Recognizing the broader implications of sustainability is crucial. Economic growth cannot be sustained at the expense of environmental degradation. As communities strive for development, they must prioritize sustainability to create a thriving ecosystem where both people and nature prosper. The benefits of embracing sustainable practices—ranging from cost savings to improved health outcomes—clearly illustrate that environmental responsibility is not just an ethical imperative but also an economic necessity.

Commitment to sustainability enhances a community's reputation, attracting businesses and residents who value environmental responsibility. As global awareness of environmental issues intensifies, areas that champion sustainable practices may become more appealing to potential investors and newcomers. A strong sustainability record can position a community as a leader, fostering opportunities for future economic development and innovation.

Ultimately, the benefits of integrating environmental sustainability with economic growth are profound and far-reaching. By actively working toward sustainability, communities can create an environment where residents thrive, economies flourish, and nature is nurtured. Through collective action, informed decision-making, and a commitment to responsible practices, communities can not only navigate the challenges of modern life but also pave the way for a healthier, more sustainable future. This journey requires vision, collaboration, and resilience, but the rewards—both environmental and economic—are well worth the effort.

Challenges in Implementing Smart Systems

Implementing smart systems presents a complex landscape of challenges that can hinder progress, even in communities eager to embrace innovation. While these systems promise improvements in efficiency, quality of life, and operational effectiveness, the journey toward realizing these benefits often encounters significant obstacles. Understanding these challenges is crucial for successfully navigating the implementation process and maximizing the potential of smart technology.

One of the primary challenges in adopting smart systems lies in the technological infrastructure. Many communities, particularly those in underserved or rural areas, may lack the necessary infrastructure to support advanced technologies. High-speed internet access is a fundamental requirement for many smart systems, enabling data transmission and real-time communication. Without a reliable internet connection, smart sensors, devices, and networks cannot function optimally. This digital divide can result in unequal access to technology, further exacerbating existing inequalities within communities. Addressing this infrastructure gap is essential for every resident to benefit from smart systems.

Financial constraints also pose a substantial hurdle. While the long-term benefits of smart systems can be significant, the initial investment often requires substantial financial resources. Budgeting for new technologies can be daunting for local governments and businesses, especially when competing priorities, such as education and healthcare, vie for limited funds. Striking a balance between immediate needs and future investments becomes a critical challenge. Moreover, organizations must consider ongoing operational costs, including maintenance and training, which can further strain budgets. Developing a clear business case that outlines both short-term costs and long-term savings can facilitate discussions on funding these initiatives.

Resistance to change is another common obstacle. Individuals and organizations accustomed to traditional practices may be hesitant to embrace new technologies. Fear of the unknown, concerns about job displacement, or simply a preference for familiar methods can foster skepticism and impede progress. Engaging stakeholders from the outset and involving them in the decision-making process can mitigate resistance. Providing clear communication about the benefits of smart systems—enhanced efficiency, improved safety, and increased productivity—can help alleviate fears and foster a culture of innovation.

Data security and privacy concerns are increasingly critical considerations as smart systems rely on vast amounts of data to function effectively. The collection and sharing of sensitive information raise alarms for many individuals. Communities must ensure transparency about how data is collected, stored, and utilized. Establishing comprehensive data governance policies is essential to protect personal information and build trust among residents. Investing in robust cybersecurity measures is also necessary to safeguard systems from potential breaches that could jeopardize sensitive data and undermine confidence in the technology.

Integrating smart systems with existing technologies can be a significant challenge. Many organizations already use legacy systems that may not easily interface with newer technologies. This incompatibility can lead to increased costs and complexity during implementation. Successful integration often requires considerable IT resources and expertise, which may be lacking in smaller organizations. To navigate this challenge, organizations must conduct thorough assessments of existing technologies and develop clear strategies for integration. A step-by-step approach, Gradually phasing in new systems can ease the transition and reduce resistance from staff and stakeholders.

Training and capacity building are vital components of successful smart system implementation. Users must be equipped with the skills and knowledge necessary to operate new technologies effectively. Concerns around the learning curve can deter staff from engaging with the systems. Creating a supportive training environment, with resources tailored to varying levels of technological proficiency, can enhance user confidence. Ongoing support mechanisms, including help desks and user guides, foster a culture of learning that is crucial for long-term success.

The complexity of smart systems can lead to operational challenges. These systems often rely on intricate algorithms and machine learning processes that can be difficult to understand. Misinterpretation of data or administrative errors can lead to incorrect decision-making, resulting in unintended consequences. Implementing clear protocols and regular audits can help mitigate these risks. Organizations must prioritize continuous monitoring and evaluation processes to assess system performance and refine methodologies as needed.

Another challenge arises from potential regulatory barriers. Emerging technologies frequently find themselves outpacing existing regulatory frameworks designed to govern conventional systems. This mismatch can create uncertainty, resulting in delayed implementation. Engaging with policymakers early in the process can help shape regulations that support innovation while still ensuring public safety. Collaborating with industry partners and stakeholders can facilitate a better understanding of the implications of smart systems and contribute to the development of relevant policies.

Community engagement plays a pivotal role in the successful implementation of smart systems. Absent inclusive engagement practices, initiatives may fail to address the specific needs and preferences of community members. Residents need to understand how these systems will impact their lives. Conducting surveys, holding public forums, and creating opportunities for dialogue can encourage participation and feedback. Listening to resident concerns and adapting solutions to align with community values fosters a sense of ownership and investment in the success of the project.

Establishing clear performance metrics is crucial for evaluating the effectiveness of smart systems. Without a defined framework for measurement, organizations may struggle to assess progress, identify areas for improvement, or justify the investment in these technologies. Developing comprehensive evaluation criteria that take user experience, cost savings, and operational efficiency into account can provide valuable insights to guide future decisions. Regularly reviewing performance data enables organizations to remain adaptable and responsive to changing community needs.

The volatility of technology is another significant concern. Rapid advancements mean that today's state-of-the-art systems can quickly become outdated. Organizations must remain vigilant about ongoing developments, updates, and best practices within the field. Developing partnerships with technology providers can facilitate access to the latest innovations and ensure systems remain relevant and effective. Fostering a culture of adaptability within organizations will prepare staff to embrace changes and evolving technologies, rather than fearing them.

Addressing environmental sustainability is increasingly essential in the discussion of smart system implementation. While technology can drive efficiencies, organizations must also consider the environmental footprint of their systems. Sustainable practices must be embedded into the planning and operational processes. Ensuring that technologies are energy-efficient and environmentally friendly demonstrates a commitment to not only economic growth but also ecological responsibility. Pursuing green certifications and working with environmentally conscious vendors can amplify the impact of these initiatives.

Navigating these multifaceted challenges requires strong leadership and collaboration across various sectors. Leaders must champion the vision to implement smart systems while fostering a culture of innovation and resilience. Building coalitions involving municipal governments, businesses, and community organizations can generate support, share resources, and leverage diverse perspectives. A united front will increase visibility and drive progress towards the effective integration of smart systems.

While the challenges in implementing smart systems are substantial, they are not insurmountable. With a proactive approach and thoughtful strategies, communities can create solutions that not only address these obstacles but also pave the way for a more innovative and sustainable future. As society advances into an increasingly interconnected world, embracing these technologies becomes essential for enhancing the quality of life, improving economic prospects, and fostering resilience in the face of change. The journey may be complex, but the potential rewards offer a compelling vision of what is possible when communities rise to the challenge of transformation.

Chapter 2: Technological Innovations in Waste Collection

IoT-Enabled Waste Bins and Sensors

The integration of IoT (Internet of Things) technology into waste management is revolutionizing how communities handle their refuse. IoT-enabled waste bins equipped with sensors offer real-time data on fill levels, optimizing collection processes, reducing costs, and promoting sustainable practices. As urban areas grow and waste generation increases, these smart solutions provide a way forward, tackling issues of inefficiency and environmental impact.

Imagine a bustling city where garbage collection routes are optimized dynamically based on real-time data. Traditional waste management relies on fixed schedules that often lead to half-full bins being emptied, wasting resources and time. Conversely, IoT-enabled bins equipped with sensors can monitor waste levels and transmit this information to a central management system. This allows waste collection services to dispatch trucks only when bins are nearing capacity, resulting in reduced fuel consumption and lower operational costs.

The core component of these smart waste bins is the sensor technology they employ. Typically, these sensors detect fill levels and can also monitor other metrics, such as temperature and weight. By transmitting this data via cellular or Wi-Fi networks, waste management systems gain invaluable insights into waste generation patterns. This data can help cities and municipalities plan more efficient collection routes and schedules.

Implementing IoT-enabled waste bins requires careful planning and execution. First, stakeholder engagement is essential. Collaborating with local governments, waste management companies, and community members helps identify specific needs and concerns. Addressing these from the outset ensures that the solution aligns with community objectives, fostering acceptance and enthusiasm for the technology.

Project feasibility must be assessed with tangible outcomes in mind. Initial costs can be a barrier; thus, developing a clear business model demonstrating long-term savings is crucial. This could include projections of reduced labor and fuel costs, as well as potential revenue from recycling. By outlining these financial benefits, stakeholders can justify the investment in smart waste technology.

Once funding is secured, selecting the right technology is critical. IoT sensors vary in functionality, accuracy, and cost, so choosing the appropriate model for your specific application is essential. Some sensors use ultrasound to measure waste levels, while others employ load cells to gauge weight. Moreover, the communication protocol—whether LoRa, NB-IoT, or traditional cellular—can impact both data transmission reliability and operational costs. Collaborating with technology providers who understand the waste management sector can facilitate the right choices and ensure compatibility.

After selecting the technology, the installation process must be managed effectively. Bins need to be strategically placed throughout the service area, considering foot traffic, proximity to businesses, and community usage patterns. Analyzing data from existing waste collection methods can provide insights into optimal bin placement. For instance, areas with high foot traffic may necessitate more frequent monitoring compared to quieter locations, allowing for targeted installation of IoT-enabled bins.

As these bins begin operation, ongoing maintenance and monitoring are essential. Regular checks on sensor functionality and bin condition can prevent data loss and ensure optimal performance. Establishing a routine for maintenance, addressing sensor glitches or malfunctions promptly, is vital to retaining the reliability of the systems. Furthermore, collecting data continuously allows for performance analysis, revealing patterns in waste generation and helping refine collection schedules.

The community's role in this ecosystem cannot be overlooked. Educating residents about the benefits of smart bins not only fosters engagement but also encourages responsible waste disposal practices. Awareness campaigns can highlight how IoT technology optimizes waste collection, reduces environmental impact, and potentially lowers costs for services. Involving communities in the solution—asking them to report issues with bins or participate in monitoring—can enhance accountability and encourage proactive engagement.

Data privacy and security are also significant considerations. With increasing amounts of data being collected, ensuring that residents' personal information is protected must be a top priority. Setting up data governance frameworks that outline data usage, access, and protection strategies helps alleviate concerns and builds trust between the community and waste management authorities. Regular audits and transparent reporting can further enhance accountability and security.

Sustainability is at the heart of implementing IoT-enabled waste management solutions. By streamlining waste collection processes, cities can reduce greenhouse gas emissions associated with garbage trucks. Moreover, sensors that identify recyclable materials can help municipalities improve sorting processes at recycling facilities. Understanding what waste is generated can lead to better recycling programs and initiatives aimed at reducing total waste output.

In addition to improving collection efficiency, smart bins can help reduce litter in public spaces. Equipped with sensors that alert sanitation crews when a bin is full, these systems ensure that waste is promptly removed, keeping neighborhoods clean and appealing. The visual presence of smart bins can also encourage responsible disposal, as communities recognize that their waste is being monitored systematically.

Engaging with emerging technologies can enhance the capabilities of IoT-enabled waste systems further. For instance, integrating AI analytics can help forecast waste generation trends based on historical data, local events, and seasonal fluctuations. These insights empower cities to proactively manage their waste systems, adjusting collection frequencies and optimizing resource allocation.

The role of partnerships in this ecosystem cannot be underestimated. Collaborating with technology firms, academic institutions, and other municipalities can provide innovative solutions and shared best practices. Joint initiatives enable resource sharing, creating a community of practice that fosters collaboration and knowledge exchange, ultimately enhancing the effectiveness of waste management technologies.

As smart bins become more commonplace, the potential for future developments in this field is immense. Combining IoT technology with other smart city projects—like traffic management or environmental monitoring—can yield comprehensive urban solutions. This interconnected approach can lead to smarter, more resilient cities, equipped to handle the complexities of modern urban life.

By embracing IoT-enabled waste management systems, cities and communities can optimize their operations, respond to environmental challenges, and enhance quality of life for residents. The journey may require significant effort, collaboration, and investment, but the rewards—a cleaner environment, improved efficiency, and greater community engagement—underscore the necessity of these innovations. As urban areas continue to grow, smart waste solutions will become an integral part of maintaining sustainability and efficiency in the cities of the future.

Automated Waste Collection Vehicles

The landscape of waste management is undergoing a radical transformation, driven by innovations in technology. Automated waste collection vehicles (AWCVs) are at the forefront of this change, revolutionizing how cities approach garbage collection. These vehicles promise not only to enhance operational efficiency but also to reduce environmental impacts and improve overall service quality. Understanding the intricacies of these automated systems, from their technology to their implementation, offers valuable insights for municipalities looking to modernize their waste management processes.

Automated waste collection vehicles rely on sophisticated technology to operate efficiently. Equipped with advanced sensors, cameras, and GPS systems, these vehicles can navigate urban environments with minimal human intervention. The operational mechanics of AWCVs combine real-time data collection with machine learning algorithms, enabling them to adjust routes dynamically based on traffic patterns, weather conditions, and waste generation trends. This adaptability ensures that collection routes are optimized for speed and efficiency, reducing fuel consumption and operational costs.

One of the primary benefits of AWCVs is their potential to enhance safety. Traditional waste collection can be hazardous. Manual collection exposes workers to traffic, lifting injuries, and harsh environmental conditions. By automating these processes, cities can significantly reduce the risks associated with garbage collection. The vehicles are designed to operate independent of human drivers, often allowing for remote or semi-autonomous control. This not only protects workers but also minimizes the chances of accidents in densely populated urban areas.

As cities consider implementing automated waste collection, stakeholder engagement becomes crucial. Involving local governments, waste management firms, and the community throughout the planning process can ensure that the technology meets the specific needs of the area. Transparent communication about the benefits and potential challenges of AWCVs fosters community buy-in, helping to address concerns related to job displacement or operational disruptions.

Financial considerations are paramount when it comes to adopting AWCV technology. The initial investment can be substantial, encompassing costs for vehicles, technology integration, and staff training. However, the long-term savings often outweigh these upfront expenditures. Cities must conduct thorough cost-benefit analyses that account for reduced labor costs, lower fuel consumption, and improved operational efficiencies. By framing the investment within the context of future savings and potential environmental benefits, stakeholders can better understand the value of transitioning to automated systems.

Selecting the right technology is another critical step. AWCVs incorporate various features, from robotic arms for bin lifting to advanced navigation systems. Different models may offer varying capabilities, whether focused on agility in urban environments or capacity for larger waste loads. Collaboration with technology providers that specialize in waste management can help municipalities identify the vehicles best suited to their unique challenges and optimal performance targets.

Deployment of these vehicles requires a strategic approach. Mapping out collection routes based on existing waste generation patterns is essential. Detailed analysis of historical data can inform decisions on where to allocate resources. For instance, areas with higher population density may benefit from more frequent pickups, while less populated zones may require fewer stops. Furthermore, integrating AWCVs with existing waste management systems allows for a smoother transition, minimizing disruptions to current operations.

Once operational, ongoing training for staff is vital. While the vehicles may require less manual intervention, operators must be well-versed in trouble-shooting, maintenance, and ensuring safe operation. Regular training sessions and workshops can prepare teams for adapting to this shift in technology. By fostering a culture of learning and adaptation, organizations can ensure that all personnel are equipped to manage and support the automated systems effectively.

Data collection should be a continuous process, feeding insights back into the waste management system. AWCVs generate vast amounts of operational data, from vehicle performance to route efficiency. By analyzing this information regularly, municipalities can identify trends, problems, and opportunities for improvement. For instance, if a certain route consistently results in delays, that data can lead to re-evaluation of the collection schedule or adjustments in route planning.

Despite the numerous advantages, challenges associated with AWCV implementation cannot be ignored. Transitioning to automated systems may lead to concerns about job displacement among waste management workers. Open dialogue and collaboration with labor organizations can help address these issues proactively. Offering reskilling opportunities for workers can facilitate their transition into new roles, whether within the expanded technological framework or in community engagement capacities focused on sustainability initiatives.

Environmental sustainability is a motivating factor behind the adoption of AWCVs. These vehicles are often designed with eco-friendly features, such as electric powertrains, reducing emissions associated with waste collection. Furthermore, by optimizing collection routes, AWCVs minimize fuel usage, contributing to lower overall carbon footprints. Municipalities committed to sustainability can showcase their efforts through the implementation of these innovative technologies, attracting support from environmentally conscious constituents.

Monitoring and evaluating the impact of automated waste collection vehicles is crucial for long-term success. Setting up key performance indicators (KPIs) allows cities to track progress and ensure alignment with their sustainability goals. Metrics might include reductions in fuel consumption, improvements in collection times, and increases in recycling rates. By regularly reviewing performance against these standards, municipalities can adjust their strategies and maximize the benefits of their automated systems.

Public engagement can further enhance the effectiveness of AWCVs. Communities that are informed about how the new systems work and the benefits they provide are more likely to embrace the changes. Educational initiatives that highlight the positive impacts of automated collection—such as improved cleanliness in public spaces and reduced environmental impacts—can foster support and cooperation from residents. Providing platforms for community feedback enables citizens to voice their concerns and suggestions, ensuring that their needs are addressed.

The future of waste management is undeniably linked to automation. As urban populations continue to grow, cities will face increasing pressures to manage waste sustainably and efficiently. Automated waste collection vehicles stand out as a viable solution, offering transformative potential for waste management practices. As the technology continues to evolve, innovations such as enhanced AI algorithms for route optimization and advanced sorting capabilities will further strengthen the role of automation in waste collection.

Collaboration with other cities experimenting with AWCVs can yield important insights and best practices. Sharing experiences on challenges faced during implementation and successes achieved can enhance the readiness of municipalities considering similar transitions. Developing networks of cities committed to smart waste management allows for exchange of knowledge and strategies, positioning all involved parties for greater success.

Final reflections on the integration of automated waste collection vehicles highlight the importance of a systematic approach—considering technology, operational efficiency, and community engagement as interconnected aspects of a broader waste management strategy. While challenges exist, the benefits of adopting such innovations are substantial. By embracing automation, cities can pave the way for a cleaner, more sustainable future, ensuring that waste management keeps pace with the needs of modern urban life. Ultimately, as technology advances, the opportunity to create smarter, more efficient communities will prevail, ushering in an era of responsible waste management that prioritizes both operational excellence and environmental stewardship.

Data-Driven Route Optimization

Efficient waste collection is a critical aspect of modern urban management, and data-driven route optimization has emerged as a powerful tool to enhance this process. By harnessing data analytics and real-time information, cities can significantly improve the efficiency of their waste collection systems, reduce operational costs, and minimize environmental impact. This chapter delves into the importance of data-driven approaches in route optimization, outlining practical steps for implementation while addressing the necessary technology, stakeholder engagement, and potential outcomes.

Start by recognizing that traditional waste collection methods often rely on fixed schedules and pre-determined routes. While this approach may appear straightforward, it often leads to inefficiencies. Bins may be half-empty during pickups, resulting in wasted resources, time, and fuel. By integrating data-driven strategies into route planning, municipalities can tailor their approaches to actual needs, ensuring that waste collection is both responsive and efficient.

At the core of data-driven route optimization is the collection and analysis of relevant data. Various types of data can be leveraged to inform decision-making, including historical waste generation statistics, traffic patterns, weather conditions, and real-time information from sensors on waste bins. This multifaceted approach allows for a more nuanced understanding of when and where waste is generated, enabling better resource allocation.

Start by laying the groundwork with foundational data. Historical data serves as an essential starting point. Collecting and analyzing past waste collection trends can reveal patterns that may not be immediately apparent. For instance, certain neighborhoods may generate more waste on weekends or during specific seasons due to events or holidays. By understanding these trends, cities can adjust their routes and schedules accordingly, ensuring they meet the fluctuating demands of different areas.

Integrating real-time data collection is crucial for ongoing optimization. Many municipalities now employ IoT-enabled waste bins equipped with sensors that monitor fill levels. These sensors provide immediate feedback on when bins are approaching capacity, allowing for more timely pickups. Rather than relying solely on fixed schedules, waste management teams can adapt routes based on actual data from individual bins. This responsiveness not only improves efficiency but also enhances customer satisfaction, as residents can expect timely service.

Next, the use of geographic information systems (GIS) plays a pivotal role in route optimization. GIS technology enables waste managers to visualize data spatially, making it easier to identify patterns and trends in waste generation across the city. By overlaying demographic data, traffic congestion maps, and historical waste statistics, decision-makers can develop optimized routes that account for multiple variables. GIS also allows for simulations to predict the outcomes of various routing scenarios, enhancing the planning process.

Implementing data-driven route optimization involves collaboration among various stakeholders. Engaging local government agencies, waste management companies, and community members is vital for success. Before deploying new technologies, it is critical to facilitate dialogue with all parties involved to address concerns and ensure alignment with community needs. Conducting workshops or informational sessions can raise awareness around the benefits of data-driven optimization, fostering support for the initiative.

Financial considerations are paramount when adopting new systems. While initial investments in technology may seem daunting, the long-term savings can be substantial. Conducting a cost-benefit analysis reveals how optimized routes can lead to reduced fuel consumption, lower labor costs, and improved operational efficiency. By framing financial investments in the context of future savings and enhanced service delivery, municipalities can better justify the transition to data-driven methodologies.

Once the decision is made to move forward with data-driven route optimization, implementing the technology becomes the next step. This may involve acquisition of sensors for waste bins, GPS tracking for collection vehicles, and software solutions to analyze the collected data. Integrating these technologies can be complex, requiring training for staff and alignment of existing systems with new tools. Collaborating with technology providers that have experience in waste management can streamline the process and help cities avoid common pitfalls.

As the system becomes operational, continuous monitoring and analytics are essential for maintaining efficiency. Waste collection teams should regularly review performance metrics to track improvements in route effectiveness. Key performance indicators (KPIs) may include average collection times, fuel consumption per route, and response times to overfilled bins. Regular analysis of these metrics allows cities to identify areas for further optimization.

Community engagement remains a cornerstone of successful implementation. As new systems are adopted, residents should be informed of changes and improvements made through data-driven routing. Educational campaigns that explain the technology and its benefits can foster understanding and support from the community. Engaging with citizens through surveys or feedback sessions can also provide valuable insights into how the service is perceived.

Another significant advantage of data-driven route optimization is its potential to contribute to sustainability efforts. Reducing unnecessary trips not only cuts costs but also lowers emissions associated with waste collection vehicles. By optimizing routes based on data, municipalities can create environmentally friendly waste management systems that align with broader sustainability goals. Incorporating electric or hybrid vehicles into the fleet further enhances this commitment to sustainability.

Future advancements in data-driven route optimization will likely revolve around the increasing sophistication of data analytics. As machine learning algorithms become more complex, they can predict waste generation patterns with greater accuracy, taking into account a wider range of variables, such as local events or socioeconomic factors. This predictive capability allows waste management systems to be proactive rather than reactive, ensuring resources are allocated effectively before waste generation spikes.

In addition to predictive analytics, incorporating citizen engagement platforms can further enhance waste collection efficiency. Communities can leverage mobile apps where residents report issues like overfilled bins, missed pickups, or specific waste collection needs. This reporting system not only improves communication between residents and waste management services but also provides additional data points for optimizing routes based on community feedback.

Ultimately, the move towards data-driven route optimization represents a significant shift in how cities manage waste collection. It embodies a proactive approach to resource management, prioritizing efficiency while simultaneously enhancing community satisfaction and environmental sustainability. By using a blend of historical data, real-time analytics, and community input, municipalities can create responsive waste management systems that are well-equipped to navigate the complexities of modern urban life.

Embracing these innovations may entail challenges, yet the benefits can far outweigh the obstacles. Cities that take the long view, investing time and resources into developing data-driven strategies, will find themselves positioned at the forefront of efficient and sustainable waste management. As urban populations continue to grow and environmental concerns mount, the need for intelligent, responsive systems will only become more pronounced. Through careful planning, stakeholder engagement, and technological adoption, municipalities can ensure they are not just keeping pace but leading the way in the evolution of waste management practices.

RFID and GPS Tracking Technologies

The advent of RFID (Radio Frequency Identification) and GPS (Global Positioning System) tracking technologies has transformed various industries, enabling greater efficiency, accountability, and accuracy in operations. These technologies have become indispensable tools for businesses and municipalities alike, offering a wealth of data and insights that nurture informed decision-making. Understanding how RFID and GPS function, their applications, and best practices for implementation can yield significant benefits.

RFID technology uses electromagnetic fields to automatically identify and track tags attached to objects. Each tag contains a microchip and an antenna, allowing it to communicate with an RFID reader. This interaction facilitates the seamless transmission of information about the tagged item, including its location and status. The beauty of RFID lies in its ability to read multiple tags simultaneously, making it especially valuable in environments where managing large quantities of inventory or assets is essential.

In contrast, GPS tracking utilizes satellite signals to determine the location of objects in real time. By receiving signals from satellites, GPS devices can pinpoint their precise locations on Earth with remarkable accuracy. This capability is not only critical for navigation but also for managing fleets, monitoring deliveries, and enhancing security by keeping track of valuable assets.

The synergy between RFID and GPS is particularly compelling. While RFID excels in tracking items in confined spaces—such as warehouses or retail stores—GPS shines in outdoor applications, offering visibility over larger areas. Combining these technologies can create a comprehensive tracking system that maximizes the strengths of both, providing end-to-end visibility whether items are in transit or stored.

When considering the adoption of RFID and GPS technologies, it is vital to assess specific business needs. Begin with a clear understanding of goals. For instance, if the primary aim is to streamline inventory management, RFID may be more relevant. Alternatively, if tracking delivery routes or fleet management is the priority, GPS would be the logical choice. Defining objectives early on will guide subsequent decisions throughout the implementation process.

A common application of RFID technology is in supply chain management. Businesses facing challenges in tracking inventory levels often find relief through RFID systems. By attaching RFID tags to products, companies can monitor stock levels in real time, reducing the risk of overstocking or stockouts. This capability not only enhances operational efficiency but also minimizes carrying costs associated with excess inventory.

Once RFID tags are integrated into operations, the data collected can yield valuable insights. When paired with analytics software, businesses can analyze patterns in inventory turnover or identify areas of theft and loss. This data-driven approach enables companies to implement targeted strategies to address inefficiencies, thereby optimizing their inventory processes.

For those leaning toward GPS tracking, the benefits in fleet management are particularly significant. GPS enables businesses to monitor the real-time location of vehicles, providing essential information for routing and scheduling. By analyzing the routes taken, companies can identify inefficiencies, such as excessive idling or detours. This analysis can lead to improved fuel efficiency and reduced operational costs.

Moreover, GPS technology enhances customer service. With real-time tracking, businesses can provide accurate delivery windows, keeping customers informed and satisfied. This transparency fosters trust and builds a positive brand reputation. As more consumers demand visibility regarding their orders, embracing GPS tracking can give companies a competitive edge.

Implementing RFID and GPS technologies requires careful planning and consideration. Key steps in successful implementation include selecting the appropriate hardware and software, establishing a robust data management strategy, and training staff on new systems. The choice of RFID tags, for instance, should be informed by the environment in which they will be used. Passive RFID tags, which require an external reader to transmit data, can be ideal for inventory, while active tags, which have their own power source, are better suited for tracking assets over longer distances.

Equally important is the software that accompanies these technologies. A comprehensive tracking system should integrate seamlessly with existing operational tools and offer user-friendly interfaces for managing and analyzing data. Selecting the right software vendors who understand the specific needs of the business is crucial for fostering effective collaboration.

Training staff on new technologies cannot be overlooked. Employees must understand how to use the systems correctly to reap maximum benefits. Building a culture of data-driven decision-making encourages workers to embrace these innovations, ultimately leading to higher adoption rates and more effective utilization of the technologies.

Security concerns also deserve attention when implementing RFID and GPS. As these systems collect and transmit sensitive data, establishing stringent security protocols is necessary to mitigate potential risks. This includes encrypting data, ensuring secure data storage, and limiting access to authorized personnel only. Regular audits of security practices can help identify vulnerabilities and enhance overall data protection.

In addition to operational improvements, the use of RFID and GPS contributes to more sustainable practices. By optimizing routes and reducing excess travel, businesses can decrease their carbon footprints. Additionally, efficient inventory management aids in minimizing waste, as organizations can manage stock levels more effectively. As sustainability continues to be a focal point for consumers, leveraging these technologies aligns with broader expectations for environmental responsibility.

The integration of RFID and GPS can produce comprehensive insights into operational performance. These insights can drive continuous improvements. Businesses should embrace a mindset of ongoing refinement, regularly evaluating the data collected to identify trends and uncover opportunities. This iterative approach can help organizations adapt to changing market conditions or customer demands.

As businesses evolve, scalability becomes a critical consideration. RFID and GPS solutions should be designed with expansion in mind; the ability to seamlessly add new tags or tracking devices can ensure that systems remain effective as needs change. Choosing systems that offer flexibility and are adaptable to growth will pay dividends over time.

The future of RFID and GPS technologies promises even greater advancements. Innovations such as blockchain integration can enhance the security and traceability of assets, allowing for more transparent supply chains. Additionally, advancements in battery technology may lead to more efficient and longer-lasting active RFID tags. As industries continue to explore these possibilities, staying informed about technological advancements will help companies leverage emerging solutions effectively.

Embracing RFID and GPS technologies is not merely about keeping pace; it represents a commitment to operational excellence. By understanding how these technologies work together to offer real-time insights, organizations can improve efficiency, enhance customer satisfaction, and drive sustainability. As competition intensifies in various sectors, those who harness the power of tracking technologies will position themselves as leaders in their fields.

In conclusion, the implementation of RFID and GPS tracking technologies can lead to transformative results across diverse environments. The benefits extend far beyond simple tracking; they encompass improved efficiency, enhanced visibility, and greater accountability. As businesses embark on this journey, a careful assessment of needs combined with strategic implementation will pave the way for successful outcomes, ensuring that organizations remain agile and well-prepared for the future.

Real-World Applications and Case Studies

The shift toward innovative waste management practices has led to the adoption of advanced technologies that reshape how cities and organizations handle their refuse. By examining real-world applications and case studies, it becomes evident that effective waste management systems are achievable through a combination of strategic planning, community engagement, and the implementation of technology. The following narratives exemplify how various municipalities and organizations have successfully navigated the complexities of waste management through innovative approaches.

Consider the city of San Francisco, which set an ambitious goal of zero waste by 2030. This initiative is not just about reducing landfill waste but transforming the entire waste management framework. Central to its success is the comprehensive use of data analytics coupled with community involvement. The city partnered with local organizations to educate residents about recycling and composting, significantly enhancing participation rates. Data collected from waste audits revealed patterns that informed service adjustments, optimizing collection routes and pickup schedules. This combination of community engagement and data-driven decision-making has positioned San Francisco as a leader in waste management innovation.

Another example comes from Seoul, South Korea, which implemented a user-pay system to encourage waste reduction. By charging residents based on the amount of waste they generate, the city incentivized recycling and composting. This program was supplemented with RFID technology, allowing for precise tracking of waste volumes and behaviors at the household level. As a result, Seoul saw a drastic decrease in waste production—by more than 30% in just a few years. Residents became more conscious of their waste habits, fostering a culture of sustainability. This model demonstrates how financial incentives, combined with technology, can effectively drive behavioral change in waste management.

Seattle presents another compelling case study with its advanced recycling program. The city deployed smart bins equipped with sensors that monitor waste levels, enabling real-time data collection about fill rates. This information allows waste management teams to optimize pickup schedules, ensuring that bins are emptied before they overflow. Not only does this reduce operational costs, but it also minimizes public nuisance and enhances the overall aesthetic of neighborhoods. The integration of technology, alongside robust public education campaigns about recycling protocols, contributed to

Seattle achieving one of the highest recycling rates in the United States.

On the corporate side, Unilever, a global leader in consumer goods, has committed to reducing waste across its supply chain. The company introduced a program called "Waste to Value," which aims to turn waste into revenue through innovative recycling methods and partnerships. In India, for example, Unilever collaborates with local waste pickers to enhance the recycling of plastic waste. By providing training and resources, the initiative not only improves workers' livelihoods but also increases the volume of recycled materials. This case illustrates how private sector involvement can significantly bolster waste management efforts while fostering community partnerships.

Exploring the challenges faced by municipalities unveils how innovation can be spurred by necessity. In Brazil, the city of Belo Horizonte encountered a critical waste crisis due to rapid urbanization and population growth. Local leaders recognized the imperative to act and developed a decentralized waste management system. Instead of relying solely on centralized facilities, the city embraced local composting and recycling initiatives that empowered communities. By engaging local citizens in the process and promoting small-scale solutions, Belo Horizonte successfully improved waste diversion rates and reduced the environmental impact of waste management. This grassroots approach highlights the importance of community ownership in sustainable practices.

Turning attention to the United Kingdom, a collaboration between the cities of Birmingham and Coventry showcases the effectiveness of intercity partnerships. Faced with growing waste management challenges, these cities developed a joint waste collection program that centralized data tracking and analysis. By pooling resources, they not only reduced costs but also improved overall service efficiency. Each city utilized shared technology platforms to track waste generation patterns, which informed joint educational campaigns aimed at reducing waste at the source. This collaboration reflects how municipalities can leverage collective strengths to enhance waste management efficacy and improve community outcomes.

The application of circular economy principles is evident in the case of Amsterdam, which is transforming its waste management strategy into a model of sustainability. The city established a framework aimed at minimizing waste and maximizing the lifecycle of materials through innovative design and recycling initiatives. A notable initiative within this framework is the "Amsterdam Circular" program, which encourages businesses to adopt circular practices by developing closed-loop systems. This shift enables the city to preserve resources and reduce dependence on virgin materials, demonstrating how policy, design, and community engagement can work in harmony to create sustainable solutions.

Community-based initiatives, such as neighborhood cleanup events, have proven effective in various locations, garnering enthusiasm and participation. For instance, in Toronto, local organizations frequently host community cleanup days, encouraging residents to take ownership of their environment. These events not only reduce litter but also foster a sense of community and civic pride. The city supports these efforts by providing resources and incentives, such as free waste disposal services on cleanup days. These grassroots movements illustrate that sustainability can begin at the community level, impacting broader municipal goals.

Investigating the role of technology in waste management, the case of Barcelona offers intriguing insights. The city employs smart waste management systems that utilize IoT (Internet of Things) sensors to monitor waste container fill levels. This data-driven approach allows for dynamic route planning for waste collection vehicles, reducing operational costs while improving service delivery. Additionally, the city has implemented a responsive feedback mechanism, enabling residents to report issues with waste services directly through an app. This two-way communication not only enhances operational efficiency but also strengthens community engagement by involving citizens in municipal decision-making.

As cities look to the future, the integration of artificial intelligence (AI) offers fascinating possibilities for waste management. For instance, a pilot project in Los Angeles harnessed AI algorithms to analyze data from various sources, including social media and city reports, to predict waste generation patterns. By anticipating spikes in waste due to events or seasonal changes, the city could allocate resources more effectively. This proactive approach illustrates how leveraging innovative technologies can lead to more responsive waste management systems.

The challenges posed by climate change necessitate that waste management systems adapt and innovate continually. Cities like Copenhagen are leading the way by focusing on holistic strategies that integrate waste management with broader urban planning. The city is committed to becoming carbon neutral by 2025, which involves rethinking its waste streams and reducing the environmental impact of landfills. Copenhagen has embraced waste-to-energy technologies that convert non-recyclable waste into energy while promoting comprehensive recycling and composting programs. This holistic perspective underscores the interconnectedness of waste management with sustainability goals across urban environments.

In summary, examining real-world applications of innovative waste management strategies reveals a tapestry of success stories, each contributing to a shared vision of sustainability. From the ambitious zero-waste goals of San Francisco to the grassroots initiatives in Belo Horizonte, these examples highlight the multifaceted nature of effective waste management. The integration of technology, community engagement, and circular economy principles forms the cornerstone of a sustainable future. As more municipalities and organizations adopt similar strategies, the potential for transformative change in waste management grows, paving the way for healthier environments and enhanced community well-being.

Chapter 3: Advanced Waste Sorting and Processing Technologies

Machine Learning and AI in Waste Sorting

The transformation of waste management practices is increasingly driven by technological advancements, particularly in machine learning and artificial intelligence (AI). These technologies are optimizing waste sorting processes, making them more efficient and environmentally friendly. By enhancing the accuracy and speed of sorting waste, machine learning and AI are helping to reduce contamination rates in recycling streams and increase overall recycling efficiency.

An essential aspect of waste sorting is the initial identification of materials. Traditional methods often rely on manual labor, leading to errors and inconsistencies. The introduction of machine learning algorithms allows for the automated classification of various waste types. For instance, computer vision technology can analyze images of waste using convolutional neural networks (CNNs), which have shown remarkable accuracy in identifying recyclable materials. These systems can be trained on vast datasets consisting of images of recyclable and non-recyclable items, enabling the algorithms to learn from their mistakes and continuously improve over time.

One compelling example is a facility in the Netherlands that employs advanced AI-driven robots for waste sorting. These robots utilize machine learning algorithms to identify and pick up different materials, such as plastics, metals, and paper, with precision. The results are striking: the system boasts a sorting accuracy of over 90%, significantly reducing contamination and increasing the amount of recyclable materials processed. This facility demonstrates that by leveraging AI, waste management operators can achieve higher efficiency rates while lowering operational costs.

The integration of machine learning also extends to optimizing the logistics of waste collection. Utilizing predictive analytics, municipalities can forecast waste generation based on factors like seasonal trends, population density, and local events. This data-driven approach allows for smarter routing of waste collection vehicles, reducing fuel consumption and emissions. A study from a major city in the United States found that implementing machine learning for route optimization led to a 15% decrease in operational costs within the first year, alongside a notable reduction in greenhouse gas emissions.

As these technologies continue to evolve, the concept of smart bins has emerged. These bins are equipped with sensors and machine learning models that can detect the volume of waste collected and its composition. By analyzing waste trends in real-time, waste management companies can adjust collection schedules, ensuring that bins are emptied before overflowing. This proactive approach minimizes public nuisance and improves service delivery, ultimately benefiting both municipalities and citizens.

Community engagement plays a crucial role in the success of these technologies. Education initiatives that inform citizens about the importance of proper waste sorting can enhance the performance of machine learning systems. When individuals understand the significance of separating recyclables from general waste, they actively participate in improving the quality of the waste stream entering processing facilities. For example, a successful campaign in a Canadian city focused on promoting recycling behaviors. As a result, contamination rates in recycling bins fell by 25%, illustrating how community involvement can amplify the impact of technological advancements.

Collaboration between technology companies and waste management organizations is also pivotal for developing effective machine learning solutions. Startups specializing in AI for waste sorting have emerged, creating bespoke systems tailored to the needs of local waste management agencies. These partnerships lead to the sharing of data and insights, fostering innovation that can be scaled across various markets. For instance, a tech firm partnered with a municipal waste service provider in Europe to develop an AI tool that predicts contaminant presence in different waste streams. As a result, the facility was able to implement targeted training programs for staff, enhancing the sorting process further.

Despite the promising advancements, challenges remain in implementing machine learning and AI solutions. One significant barrier is the initial investment required for setting up the technology, which can be daunting for smaller municipalities. However, many cities are recognizing the long-term benefits that these systems can provide, such as increased recycling rates and reduced landfill costs. Government grants and incentive programs that encourage technological adoption can make it more feasible for these municipalities to embrace machine learning for waste sorting.

Moreover, data privacy and security are paramount concerns when utilizing machine learning systems. Implementing robust cybersecurity measures to protect sensitive information is essential. Waste management operators must ensure compliance with regulations governing data protection and privacy to foster trust among citizens. It is vital to create transparent systems where individuals are informed about how their data is being used while also demonstrating the tangible benefits of these technologies to the community.

As machine learning and AI continue to advance, the potential for even greater innovations in waste sorting emerges. For instance, the concept of deep learning, a subset of machine learning, can further enhance sorting accuracy by analyzing more complex patterns in waste material classification. Researchers are exploring the use of deep learning models that can combine various data inputs, including visual, textual, and contextual data, to refine the sorting process even more. This holistic approach could revolutionize how waste is processed globally.

Looking ahead, the integration of machine learning and AI into waste sorting systems will likely play a critical role in shaping the future of sustainable waste management. As cities grapple with increasing waste generation and ever-tightening recycling targets, the need for efficient and effective sorting solutions has never been more pressing. Machine learning can facilitate this transition by delivering rapid, accurate sorting capabilities that address the challenges faced by municipal waste systems today.

In conclusion, the application of machine learning and AI in waste sorting presents a transformative opportunity for enhancing waste management practices. From automated sorting systems in state-of-the-art facilities to predictive analytics that optimize collection logistics, these technologies are reshaping the landscape of recycling and waste management. Engaging communities, fostering collaboration between technology innovators and waste management agencies, and addressing the challenges of implementation will be critical in maximizing the benefits of these advancements. By committing to integrating machine learning and AI into waste sorting practices, municipalities can pave the way toward a sustainable future, ultimately benefiting not only their communities but the environment as a whole.

Robotics and Automation in Recycling Facilities

The integration of robotics and automation in recycling facilities represents a significant leap toward achieving more efficient and effective waste management practices. As the volume of waste continues to rise globally, traditional manual sorting processes struggle to keep pace, often leading to high contamination rates and missed recycling opportunities. By implementing robotic systems, facilities can enhance their operational efficiency, reduce labor costs, and improve the quality of the recyclables they process.

Imagine walking into a modern recycling facility, where the air is filled with the sounds of machinery quietly accomplishing what once required countless hours of manual labor. Robotic arms equipped with advanced sensors and cameras analyze the materials on the conveyor belts, swiftly identifying, sorting, and separating various items—plastic, metal, paper—at remarkable speeds. This technology not only accelerates sorting processes but also minimizes the risk of human error.

In one notable case, a facility in Sweden adopted robotic sorting technology that utilizes machine vision and deep learning algorithms. These robots were programmed to recognize specific materials, allowing them to efficiently pick and separate recyclables from general waste. The results were impressive: the accuracy of sorting improved dramatically, leading to a significant increase in the volume of recovered materials. The facility reduced its contamination rates and increased its output of high-quality recyclables, illustrating how technology can drive positive outcomes in waste management.

Automation goes beyond the physical sorting of waste. It also optimizes the entire operational workflow in recycling facilities. Automated systems can manage everything from the initial collection of recyclables to compaction and baling processes. For instance, some facilities have integrated automated guided vehicles (AGVs) that transport materials within the facility without human intervention. These vehicles, using sensors and mapping technologies, navigate complex layouts to deliver items precisely where they need to go, helping streamline operations and enhance safety.

A case study involving a large U.S. waste management company highlights the benefits of integrating AGVs into their operations. By automating material transport, the company improved efficiency and reduced operational costs by more than 20%. Additionally, this shift allowed staff to focus on higher-value tasks, such as maintaining equipment and overseeing sorting processes, further elevating overall productivity.

Training and maintenance are critical components of successful integration. As facilities embrace automation, the workforce must adapt to new technologies. This shift necessitates training programs designed to upskill employees, equipping them with the knowledge and skills needed to work alongside advanced systems. Employees should feel empowered, not threatened, by automation. Facilities that prioritize continuous training foster an adaptable culture, ensuring that their teams are prepared for the evolving landscape of waste management.

It's essential to recognize that while robotics can significantly enhance efficiency, certain challenges must be addressed. The initial investment cost for robotics and automation can be substantial. However, many facilities are discovering that the long-term benefits outweigh the upfront expenditures. Implementing robotic systems can lead to a faster return on investment through increased throughput and lower labor costs. Furthermore, as technology advances and becomes more accessible, these initial costs are expected to decrease, making automation an increasingly viable option for facilities of all sizes.

Another pressing challenge is the diversity of materials processed at recycling facilities. Different types of plastics, metals, and papers require specialized handling. Robotics must adapt to these variations; thus, flexibility in design is crucial. A cutting-edge facility might employ a combination of different robotic systems, each tailored to handle specific materials effectively. This multifaceted approach ensures that the facility remains capable of processing a wide range of waste, maximizing recovery rates.

Sustainability is a growing concern in the waste management sector, and integrating robotics into recycling processes aligns with broader environmental goals. By increasing recycling rates and reducing the amount of waste sent to landfills, facilities can significantly contribute to resource conservation. Robotics further enhance sustainability by reducing energy consumption during sorting processes. Automated systems can be programmed to operate during off-peak hours, when energy costs are lower, thereby minimizing the overall carbon footprint of recycling operations.

Public perception also plays a vital role in the successful implementation of robotics in recycling. Transparency about how advanced technologies enhance recycling efforts can foster community support and encourage participation in recycling programs. When citizens see that their recyclables are being sorted more effectively, their trust in the system strengthens, leading to higher participation rates and better waste separation at the source.

An engaging narrative from a community in Canada illustrates this point. As local authorities implemented automated sorting systems and actively communicated the positive impacts of these technologies, community participation in recycling programs soared. Residents felt more confident that their efforts were making a difference, which encouraged them to be more diligent in their recycling practices. This synergy between technology and community engagement showcases how robotics can transform not just facilities, but entire recycling cultures.

Looking ahead, the field of robotics and automation in recycling is set to evolve even further. Emerging technologies, such as artificial intelligence-driven decision-making and adaptive learning systems, will enable facilities to become even smarter. For example, future systems might analyze data in real time to make immediate adjustments to sorting processes, enhancing efficiency and reducing waste on a granular level.

Furthermore, the development of increasingly sophisticated robotic designs is expected to revolutionize sorting capabilities. Innovations like soft robotics—robots designed to handle delicate materials without causing damage—could expand the range of materials that can be effectively processed. As the industry embraces research and development, a new era of automated recycling facilities may emerge, characterized by unparalleled efficiency and adaptability.

Collaboration among technology developers, operational managers, and environmental scientists will be crucial for advancing these innovations. By working together, these stakeholders can create systems that not only enhance operational efficacy but also prioritize environmental sustainability and social responsibility.

The journey toward fully automated recycling facilities is both exciting and essential. As robotics and automation continue to shape the landscape of waste management, it's apparent that the future holds immense potential. With the right investment, training, and community engagement, recycling facilities can transform into efficient hubs of resource recovery. By harnessing the power of technology, we can create a more sustainable world, where waste is truly viewed as a valuable resource rather than a burden.

Optical and Sensor-Based Sorting Techniques

The evolution of sorting technologies within recycling facilities has dramatically changed the way materials are processed and conserved. Among these innovations, optical and sensor-based sorting techniques stand out as crucial elements that improve efficiency, accuracy, and overall sustainability in waste management. These technologies harness advanced imaging systems and various sensors to identify and separate materials based on specific characteristics, revolutionizing the recycling landscape.

Imagine stepping into a modern recycling facility where sorting is not just a labor-intensive endeavor, but a highly automated process guided by sophisticated technology. Conveyor belts transport materials at high speeds, while advanced cameras and sensors scan each item in real time. The moment a plastic bottle or a metal can appears on the conveyor, the system instantly recognizes its shape, color, and material composition. This seamless processing minimizes human error and maximizes recycling rates, setting a new standard for waste management practices.

One popular optical sorting technology is near-infrared (NIR) spectroscopy. This technique leverages the unique light absorption properties of different materials to distinguish between them. When a material passes through the sensor, it is illuminated with near-infrared light. Each material reflects this light differently, allowing the sensors to identify its composition accurately. Facilities employing NIR sorting have reported significant improvements in both the quality of recyclables collected and the efficiency of the overall sorting process. For example, a facility in Germany showcased a notable increase in the recovery rates of high-value plastics, such as PET and HDPE, primarily due to the precision of NIR technology.

Another significant player in the realm of optical sorting is color recognition systems. These systems utilize high-resolution cameras that capture images of items on the conveyor belt, analyzing their color and patterns. This method is particularly effective for sorting mixed plastics, which often appear in a myriad of colors. By accurately identifying the color composition of a material, facilities can optimize their sorting strategies, ensuring that items are grouped correctly before processing. The end result is a higher-quality stream of recyclable materials that meets the stringent specifications demanded by recycling processors.

Besides vision-based technologies, various sensors supplement optical sorting systems to enhance efficiency and accuracy. For instance, electromagnetic sensors can detect metals, allowing for a more thorough separation of ferrous and non-ferrous materials. By integrating multiple types of sensors, sophisticated sorting systems can reduce contamination and improve the purity of recovered materials. This multi-sensory approach is akin to enhancing a musician's ensemble, where each instrument contributes to a more harmonious overall performance.

As these technologies continue to advance, the integration of machine learning algorithms allows for even greater enhancements in sorting precision. By continuously learning from the data collected during sorting processes, these algorithms can adapt and improve their capabilities over time. This leads to fewer incorrect identifications and better overall recovery rates. A facility that adopted such an adaptive system reported a reduction in contamination levels by over 30%, highlighting the transformative impact of combining sensor technology with advanced data analysis.

The advantages of optical and sensor-based sorting techniques extend well beyond mere efficiency. These technologies contribute to broader environmental goals by increasing recycling rates and reducing the amount of waste that ends up in landfills. The greater purity of sorted materials means that recycling facilities can produce higher-quality feedstock for manufacturers, encouraging a circular economy where materials are continuously reused.

Community engagement and education also play an essential role in the successful implementation of these sorting systems. When citizens understand how optical and sensor technologies improve recycling outcomes, they are more likely to participate actively in recycling programs. Schools, local organizations, and municipalities can collaborate to create educational campaigns that highlight these technologies' benefits, building trust in the recycling process and fostering responsible waste disposal behaviors. Such initiatives can lead to cleaner recycling streams at the source, making the sorting process even more efficient.

Despite the advantages, the transition to optical and sensor-based sorting techniques is not without challenges. The initial investment costs can be substantial, creating barriers for smaller facilities. However, as technology continues to evolve and become more affordable, the return on investment is likely to improve. Many recycling operations find that the long-term savings in labor costs, increased recovery rates, and reduced contamination levels more than justify the upfront expenses. Over time, these investments create a ripple effect, benefiting not just the facilities but also the communities they serve.

Moreover, ongoing maintenance and employee training are critical for the success of these systems. As technologies advance, operators must be prepared to manage more complex equipment and troubleshoot issues effectively. Facilities that prioritize training and maintenance not only ensure the longevity of their systems but also empower their workforce to embrace innovation, helping them adapt to an ever-changing industry landscape.

Looking forward, the future of sorting technologies promises even greater innovations. Research and development efforts are exploring the use of hyperspectral imaging, which captures a broader spectrum of light than conventional optical sorting systems. This technology could substantially enhance the ability to identify materials, allowing facilities to sort even more complex waste streams with high precision. Additionally, the integration of robotics with optical sorting technologies paves the way for even greater automation in recycling facilities, creating smarter, more efficient systems capable of handling diverse materials.

As these changes unfold, it is essential for industry stakeholders to collaborate closely. Partnerships between technology developers, recycling operators, and environmental organizations can drive innovation while ensuring that the benefits of these technologies reach their full potential. Creating a shared vision for the future of recycling will foster an ecosystem where sorting technologies play a leading role in promoting a sustainable and responsible approach to waste management.

Embracing optical and sensor-based sorting techniques not only revolutionizes the recycling process but also reinforces the message that waste can be a valuable resource. Every plastic bottle, metal can, or piece of paper represents an opportunity for recovery and reuse. By adopting advanced sorting technologies, facilities are not just improving their bottom line; they are contributing to a more sustainable future for our planet. As the world continues to grapple with the challenges of waste management, these technologies provide a beacon of hope, demonstrating that innovation can pave the way for meaningful change. In the journey toward a circular economy, optical and sensor-based sorting technologies are pivotal in transforming how we view and handle waste, making recycling smarter, cleaner, and more effective than ever before.

Enhancing Efficiency with Big Data Analytics

The landscape of recycling and waste management is undergoing a revolutionary shift, thanks to the advent of big data analytics. As facilities grapple with increasing volumes of waste and the pressing need for sustainability, data-driven strategies have emerged as vital tools in enhancing efficiency. By leveraging the vast amounts of information generated at every stage of the recycling process, operators can make informed decisions that lead to improved recycling rates, reduced costs, and a more sustainable future.

Imagine a bustling recycling facility where every piece of data—from the types of materials processed to operational downtime—is meticulously collected and analyzed. Each moment on the conveyor belt generates valuable insights that, when aggregated, tell a compelling story. Big data analytics empowers facility managers to identify patterns, predict outcomes, and make strategic adjustments that optimize operations. For instance, using data to track peak processing times can help facilities allocate resources more effectively, ensuring that staffing levels align with demand.

A significant advantage of big data analytics is its ability to enhance sorting processes. Traditional recycling operations often struggle with contamination, where non-recyclable materials inadvertently mix with recyclables. By analyzing historical data on contamination rates, facilities can identify common issues and develop targeted strategies. For example, if analysis reveals that a particular type of packaging frequently clogs the sorting line, operators can implement educational campaigns aimed at consumers, helping them understand which materials can and cannot be recycled. This proactive approach not only improves sorting efficiency but also fosters community engagement.

Moreover, predictive analytics can help facilities anticipate maintenance needs, reducing unexpected downtimes. By monitoring the performance of sorting equipment and conveyor belts, operators can predict when maintenance is required, thus preventing costly breakdowns. For instance, if data analytics indicates a gradual decline in the efficiency of a particular conveyor system, timely intervention can be scheduled. This practice enhances not only equipment longevity but also the facility's overall operational efficiency.

Integrating Internet of Things (IoT) devices into waste management processes further amplifies the power of big data analytics. IoT sensors can track the status of equipment, the flow of materials, and even environmental conditions within the facility. This real-time data stream can provide insights that were previously unattainable. For example, sensors placed on balers can monitor compaction levels and adjust settings automatically to optimize performance. The more data collected, the finer the adjustments can be, creating a responsive system that maximizes throughput.

Statistics underscore the transformative potential of big data in recycling. A study from a forward-thinking facility in California illustrated that by using data analytics to optimize sorting processes, they increased their recovery rate of valuable materials by 15% within just a year. This increase not only represented higher revenue from recovered recyclables but also contributed to reducing the carbon footprint associated with producing new materials. Such results demonstrate that enhancing efficiency through big data is not merely theoretical; it translates into tangible benefits for both the facility and the environment.

With the wealth of data available, effective visualization tools play a crucial role in data interpretation. Dashboards that present real-time analytics in an easily digestible format enable decision-makers to quickly grasp essential information. For instance, a facility manager can monitor key performance indicators, such as the volume of materials processed per hour or the percentage of contaminants detected. These visual tools empower managers to make swift adjustments based on current conditions, improving response times and facility adaptability.

Collaboration among stakeholders is another vital aspect of optimizing big data analytics in recycling operations. By sharing data across various departments—operations, maintenance, and marketing—facilities can develop integrated strategies that enhance overall performance. For example, insights from the operations team can inform marketing efforts aimed at increasing consumer participation in recycling programs, directly impacting the quality of materials collected. This holistic approach fosters a culture of continuous improvement, where every team member plays a role in advancing the facility's goals.

However, the integration of big data analytics is not without its challenges. Ensuring data quality is paramount; data that is inaccurate or incomplete can lead to misguided decisions. Establishing robust data governance practices is essential for maintaining accuracy. Regular audits of data collection processes and implementing standards for data entry can help mitigate risks associated with poor quality data. The investment in training staff to understand the importance of data accuracy pays off by ensuring reliable insights that drive efficiency.

Privacy and security concerns also arise when collecting and analyzing data. Facilities must navigate the complexities of data handling while complying with regulations. Implementing strong cybersecurity measures, such as encryption and access controls, will protect sensitive information and foster trust among stakeholders. Transparency regarding data usage is equally important; engaging with the public about how data is collected and used for optimizing recycling processes can enhance community support and participation.

As the industry progresses, embracing emerging technologies will further amplify the benefits of big data analytics. Machine learning algorithms can analyze vast datasets to derive insights that humans might overlook. By identifying intricate relationships within the data, these algorithms can suggest optimal sorting methods or predict fluctuations in recycling rates based on external factors such as seasonal trends or changes in consumer behavior. Such predictive capabilities enable proactive management, transforming how recycling facilities operate in response to evolving conditions.

Additionally, exploring partnerships with tech companies can open new avenues for innovation. Collaborations that leverage expertise in data analytics and machine learning can result in customized solutions tailored to individual facility needs. These partnerships can provide access to cutting-edge tools that enhance data collection, visualization, and predictive capabilities, ensuring that facilities remain at the forefront of recycling technology.

The future of recycling is inextricably linked to how effectively organizations harness the power of big data analytics. As environmental concerns grow, the imperative to increase efficiency in waste management becomes more pronounced. By transforming raw data into actionable insights, recycling facilities can not only improve operational performance but also contribute to a more sustainable planet. The journey is not simply about processing waste; it is about fostering a culture of innovation, accountability, and community engagement.

In this evolving landscape, recycling facilities equipped with big data analytics are not just participants in the industry; they become leaders driving change. Every decision informed by data has the potential to ripple outward, enhancing overall efficiency and benefiting society at large. This commitment to leveraging data paves the way for a future where recycling systems are smart, adaptive, and integral to achieving a circular economy. As we stand on the cusp of this new era, one thing is clear: the power of data will be a catalyst for transformative change in recycling—heralding an age of efficiency, sustainability, and

Innovations in Material Recovery Facilities

The recycling industry is experiencing a surge of innovations, reshaping how material recovery facilities (MRFs) operate. As the volume of waste increases and the demand for high-quality recycled materials grows, MRFs must adapt by integrating advanced technologies and processes. This shift not only maximizes efficiency but also enhances the overall effectiveness of recycling efforts.

At the heart of these innovations lies automation, which transforms traditional sorting processes. Many MRFs are now incorporating robotics to handle materials with speed and precision. These robotic systems use artificial vision and machine learning algorithms to identify different types of materials on the conveyor belt, enabling them to sort items accurately. For instance, a facility in Ohio recently implemented robotic arms that can sort plastics based on their shape and size, significantly increasing the rate of accurate sorting while reducing contamination levels. The sheer speed at which these machines operate allows MRFs to process greater volumes of waste with fewer human resources, driving efficiency and profitability.

Another groundbreaking technology gaining traction is optical sorting. This method employs advanced cameras and sensors to analyze materials as they pass along the conveyor belt. The machines can identify materials based on color, size, and composition, elevating the purity of sorted streams. Optical sorters are particularly effective for multi-material containers, where diverse materials require careful separation. Facilities utilizing optical sorting have seen substantial improvements in the quality of their marketed recyclables, making them more attractive to manufacturers looking to source high-grade materials.

The implementation of near-infrared (NIR) spectroscopy further enhances the capabilities of optical sorting systems. This technology allows facilities to distinguish between different types of plastics and detect labels or films that may traditionally contaminate recycling streams. NIR sensors can identify materials that are not easily distinguishable by sight, ensuring that even subtle variations in composition are accounted for in the sorting process. Facilities employing NIR sorting report significant increases in recyclable recovery rates, illustrating its effectiveness in material identification.

Integrating data analytics into the operations of MRFs is now essential for optimizing workflows. Facilities are increasingly utilizing big data to track the flow of materials, identify trends, and refine processes. For example, by collecting and analyzing data on the composition of incoming waste streams, operators can adjust their sorting techniques in real-time. Having up-to-date insights allows MRFs to preemptively address issues like equipment malfunction or adjustments in staffing needs during peak operational hours. Additionally, advanced analytics can forecast the types and quantities of materials likely to arrive at the facility, helping managers to allocate resources more efficiently.

Energy efficiency is another critical focus of innovation in MRFs. As environmental concerns mount, many facilities are looking into renewable energy sources to power their operations. Solar panels installed on facility rooftops or wind turbines can significantly offset energy costs while reducing the carbon footprint. Some forward-thinking facilities have even implemented energy recovery systems that harness waste heat produced by machinery, converting it into usable energy for operations. This not only lowers energy expenses but also promotes a more sustainable approach to material recovery.

Moreover, the design and layout of MRFs are evolving to enhance operational flow. Facilities are rethinking how processes are structured to eliminate bottlenecks and improve sorting efficiency. Modern facilities often incorporate open layouts with clear pathways for vehicles and pedestrians, allowing for smoother movement of materials. Additionally, ergonomic design principles have become a priority, ensuring that employees can work effectively and safely alongside automated systems. Enhanced worker conditions not only promote safety but also improve overall morale and productivity.

Local communities play a vital role in the success of MRFs, making public engagement a crucial aspect of innovation. Educational initiatives aimed at raising awareness about recycling practices can significantly impact the quality of materials delivered to MRFs. By informing residents about proper sorting techniques and the importance of reducing contamination, facilities can improve recycling rates right from the source. Some MRFs have implemented community outreach programs, hosting workshops and events to engage the public directly. This fosters a culture of recycling, encouraging residents to view their contributions as essential to the overall success of the facility.

Collaboration within the recycling sector also represents a significant trend. MRFs are increasingly partnering with manufacturers to create closed-loop systems where recycled materials are reintroduced into production processes. Such collaborations can streamline material flows and ensure that the recyclables sorted are of high quality, meeting the specific needs of manufacturers. For instance, partnerships with local companies have allowed MRFs to develop specialized material streams, thus enhancing the attractiveness of recycled products and expanding market opportunities.

The role of regulatory frameworks cannot be overlooked in driving innovation within MRFs. Policymakers are acknowledging the need for enhanced recycling efforts and are starting to introduce standards that encourage the adoption of advanced technologies. Incentives for investments in modern sorting equipment or tax breaks for energy-efficient upgrades can motivate facilities to innovate. Compliance with increasingly stringent recycled content mandates can also spur MRFs to enhance their processing capabilities, ensuring they can deliver materials that meet market demands.

Looking toward the future, the innovations in material recovery facilities will continue to evolve, driven by the need for increased efficiency and sustainability. Emerging technologies like artificial intelligence will play an even greater role in optimizing operations. AI algorithms capable of analyzing vast datasets from various operational aspects can suggest best practices, improve sorting accuracy, and enhance decision-making processes in real-time. The continuous evolution in technology ensures that MRFs remain agile and capable of adapting to changing market needs.

The importance of workforce training cannot be overstated as these innovations become more prevalent. Facilities implementing new technologies must invest in training their personnel to work alongside advanced machinery and systems. Continuous learning opportunities not only enhance employee skill sets but also reinforce safety practices in increasingly automated environments. A well-trained workforce is more equipped to embrace change, ensuring that the transition to innovative processes is smooth and effective.

As recycling continues to be a focal point in global sustainability efforts, the innovations permeating material recovery facilities are indispensable. They are not just about improving efficiency or profitability; they are about reimagining how we view waste. By enhancing the way materials are sorted, processed, and marketed, MRFs can contribute significantly to a circular economy where waste is minimized, and resources are continuously reused. This ambitious vision of waste management is not only achievable but essential for creating a sustainable future.

With advancements in automation, data analytics, energy efficiency, community engagement, and collaboration, MRFs are setting new standards in the recycling industry. As these facilities embrace change, they are proving that with the right innovations, recycling can be both a practical solution and a powerful movement toward environmental responsibility. The journey ahead is filled with possibilities, and with each innovation, the recycling sector inches closer to achieving its goal of a cleaner, greener planet.

Chapter 4: Digital Platforms and Waste Management Apps

Waste Management Apps for Households and Businesses

Every year, millions of tons of waste are generated globally, and managing that waste effectively is crucial for a sustainable future. In today's digital age, waste management apps have emerged as powerful tools that assist both households and businesses in organizing, reducing, and recycling their waste. These applications provide practical solutions, allowing users to track their waste, understand recycling guidelines, and adopt eco-friendly practices. The integration of technology into waste management not only simplifies processes but also promotes environmental awareness.

Using a waste management app can start with a simple download. Most of these apps are user-friendly and accessible on various devices, making it easy for anyone to engage. Once installed, users can enter their details, such as location and types of waste generated, which helps the app provide tailored information and resources. For households, this means receiving alerts about local recycling schedules and waste collection days. Understanding these schedules is vital; many people forget or mismanage their disposals, leading to missed pick-ups and overflowing bins.

Many apps go beyond basic waste tracking. They often feature informative resources on what can and cannot be recycled. For example, confusion around recycling plastics is common—different types have different requirements. A good app will educate users about local recycling rules, helping them sort their waste correctly and reduce contamination rates. Sometimes, all it takes is a quick lookup on an app to know whether to toss a container into the recycling bin or the regular trash.

For businesses, the stakes can be higher. Effective waste management can lead to significant cost savings. Many apps designed for commercial use help companies analyze their waste generation and identify areas for reduction. By inputting data on the volume and type of waste produced, businesses can develop tailored waste reduction strategies. Such insights often lead to changing procurement practices or improving operational efficiencies that reduce waste generation. Over time, this not only saves money but can also enhance a business's reputation as a responsible entity committed to environmental stewardship.

One notable feature in many waste management apps is the ability to set recycling goals. Users can establish personal targets for reducing waste, whether it's cutting down on single-use plastics or increasing recycling rates. This goal-setting feature can be motivating, turning waste management into a more engaging and rewarding endeavor. Users can track their progress over time, share their achievements with friends or colleagues, and compete in friendly challenges. Such community aspects can create a sense of accountability and a shared purpose, contributing to more sustainable habits.

Some apps even integrate with local waste disposal systems, offering real-time updates on collection services. If a collection day changes or if there's a delay due to weather, users receive notifications immediately. This means less frustration and cleaner neighborhoods. Businesses that rely on timely waste removal can benefit significantly; knowing when their bins will be emptied helps them manage space and resources more effectively.

There are also apps that facilitate the recycling of specific items like electronics or hazardous materials. For instance, instead of throwing away old electronics, users can find local recycling events or drop-off locations through their app. This is especially crucial for certain materials that require special handling. Many cities have specific guidelines for disposing of items like batteries or e-waste, and knowing how to properly recycle these can prevent environmental harm.

Furthermore, many applications offer educational content and tips on sustainability. These resources can include articles or videos on composting, upcycling, and other eco-friendly practices. For households, having access to easy how-to guides can inspire families to embrace greener lifestyles. Those running businesses can find strategies to enhance their sustainability practices, potentially leading to certifications that highlight their commitment to the environment.

Collaboration is another powerful aspect of these apps. Some platforms allow users to connect with their local community to share resources. For example, if someone has items they no longer need, they can post them on the app, offering them for free to others instead of discarding them. This sharing economy approach fosters a sense of community while reducing waste. Homes and businesses alike can benefit from initiatives that prioritize reusing items, whether through community swaps or donation drives.

Security and privacy are valid concerns in the digital realm, but many waste management apps address these issues transparently. Users should look for apps that emphasize data protection and provide clear privacy policies. Reputable applications will not share personal information without consent, ensuring that the experience remains safe while still offering valuable resources.

The evolution of technology means that many waste management apps now employ gamification strategies to make engaging with waste reduction more exciting. Users can earn points for recycling correctly, completing educational tasks, or pledging to reduce waste. Some applications even host leaderboards that highlight top users or communities, creating friendly competition that encourages proactive engagement with waste management.

For businesses, adopting a waste management application can also serve as a part of corporate social responsibility (CSR) initiatives. Many companies are expected to report on their sustainability metrics, and having accurate data from a waste management app can support transparency in these efforts. Sharing goals, achievements, and progress not only enhances accountability but also builds trust with customers and stakeholders.

As we move toward a more sustainable future, the impact of waste management apps is increasingly significant. They empower individuals and organizations to take responsibility for their waste, making it easier to adopt environmentally friendly practices. By providing personalized resources, education, and community support, these applications are essential tools in the journey towards reducing waste and promoting recycling.

The future holds great promise for the development of even more sophisticated waste management solutions. Innovations might include enhanced features like artificial intelligence to predict waste trends, or augmented reality tools that help users visualize their progress towards sustainability goals. As technology continues to advance, the potential for these apps to foster positive environmental change is limitless.

Every action counts, and with the help of waste management apps, everyone—from individual households to large corporations—can contribute to a cleaner, healthier planet. Adopting these tools not only simplifies the journey toward improved waste practices but also instills a sense of collective responsibility. Embracing sustainability is not just an individual effort; it is a community endeavor, and waste management apps are vital in guiding this transformative experience.

Online Marketplaces for Recyclables

The rise of technology has transformed virtually every aspect of our lives, and waste management is no exception. Online marketplaces for recyclables offer a promising solution, connecting those looking to dispose of materials responsibly with buyers eager to repurpose or recycle items. This shift not only promotes sustainable practices but also empowers both individuals and businesses to engage in a circular economy, where waste is minimized, and resources are reused.

Navigating these online platforms can feel daunting at first, but the key is understanding how they operate and the value they bring. Initially, users must identify the types of recyclables they wish to sell or give away. Common materials include metals, plastics, paper, and electronics, all of which have different marketplaces tailored to their reuse or recycling potential. Each platform typically specializes in specific materials, so it's essential to choose the right one for the items on hand.

For households, utilizing these marketplaces presents an opportunity to declutter while responsibly managing waste. Instead of tossing unused items into landfills, individuals can list them for sale or donation to others who might need them. For instance, a family with old electronics can use an online platform to connect with local buyers or organizations that specialize in refurbishing or recycling these gadgets. Such actions not only keep harmful materials out of landfills but also extend the life cycle of usable products.

Businesses, too, stand to benefit significantly from online marketplaces. Companies that produce significant amounts of scrap or surplus materials can find new revenue streams by selling their recyclables directly to buyers. For instance, a manufacturing firm might generate leftover metals that could be sold to metal recyclers, turning what was once considered waste into profit. This practice not only reduces disposal costs but also enhances the company's sustainability profile, aligning with broader environmental goals.

A significant advantage of online marketplaces is their expansive reach. Unlike local recycling centers with set processing limits, these platforms facilitate connections on a global scale, allowing users to find specific buyers for niche materials. This expanded marketplace means that more materials can be recycled or reused effectively. When consumers have access to a larger pool of potential buyers, it increases the likelihood that their recyclables will be put to good use, rather than ending up in a landfill.

Leveraging social media as part of this process can enhance the visibility of listings. Many people turn to platforms like Facebook Marketplace or specialized groups within these networks to advertise their recyclables. Sharing posts can generate interest and increase the chances of selling or donating items swiftly. It often takes only a few clicks to reach thousands of potential buyers within a community, amplifying the benefits of sustainability.

Transparency is another critical element in these online transactions. Reliable marketplaces typically provide extensive information about the recycling process, ensuring that users understand how their materials will be handled once sold. This transparency fosters trust between sellers and buyers, instilling confidence in the marketplace. Users can read reviews and ratings, helping them make informed decisions about where to sell or donate their materials, thus enhancing the overall user experience.

To maximize success in online marketplaces, sellers should ensure that their listings are well-crafted and detailed. Clear photographs of the items being sold, accompanied by concise descriptions, help potential buyers assess the value of recyclables accurately. Specifying the condition of the item, any relevant details, and the asking price allows for informed decisions, reducing the likelihood of misunderstandings. A thoughtfully composed listing can greatly enhance the chances of quick sales or donations.

Additionally, sellers should be aware of local regulations regarding the sale and disposal of specific materials. Certain items may have legal restrictions around their sale or require certifications to ensure responsible handling, particularly in the case of hazardous materials. Educating oneself about these regulations prevents future complications and fosters conscientious disposal practices.

Moreover, online marketplaces often include features for tracking the environmental impact of transactions. Some platforms provide users with insights into how their sales contribute to reducing waste, offering metrics on how much material has been diverted from landfills. This feedback reinforces positive behavior, encouraging users to continue participating in sustainable practices. For those looking to make a defined impact, tracking these results can motivate further engagement in recycling efforts.

Engaging in these online marketplaces not only benefits individuals and businesses but also contributes positively to communities. Increased accessibility to reusable materials helps community members source what they need locally rather than purchasing new items from major retailers. This practice strengthens local economies and reduces environmental impacts associated with shipping products across long distances.

For nonprofits, online marketplaces serve as an invaluable tool for fundraising and resource management. Organizations can collect recyclables from the community, sell them through these platforms, and use the proceeds to support their missions. This model creates a symbiotic relationship between garbage reduction and community service, ultimately leading to a greater social impact.

While online marketplaces present countless opportunities, challenges exist that users should navigate carefully. Scams and fraudulent listings are potential risks in any online transaction setting. To mitigate these risks, sellers should always conduct transactions within recognized and reputable platforms, avoiding direct transactions with strangers whenever possible. Engaging in secure payment methods further protects users from potential scams.

Building a community around recycling can enhance the overall impact of these marketplaces. Users can join networks or groups focusing on sustainability, where they can share experiences, tips, and resources related to responsible recycling. Such collaboration encourages collective action, magnifying the positive effects of participating in these online platforms.

As our world continues to grapple with growing waste issues, the importance of engaging with online marketplaces becomes increasingly evident. These tools not only enable better waste management practices at the individual level but also inspire broader societal change. Each action contributes to a larger movement toward reducing reliance on landfill disposal, fostering a culture of sustainability.

As technology evolves, these marketplaces are likely to become even more sophisticated. Innovations such as enhanced search algorithms, better user interfaces, and streamlined payment systems will improve accessibility and usability. The future holds the promise of even further collaboration between different stakeholders, creating a comprehensive network that supports sustainable waste management and recycling efforts.

Engaging with online marketplaces for recyclables replaces the narrative of convenience and disposability with one of responsibility and community engagement. Every small action counts, and by harnessing the power of these digital platforms, individuals and businesses can play a significant role in the movement toward a more sustainable future. Each recyclable item sold or donated is a step toward reducing environmental impact, building a circular economy that benefits everyone involved. Make the choice to embrace these tools, contribute to positive change, and inspire others to join in advocating for a cleaner, greener world.

Citizen Engagement through Digital Platforms

Digital platforms have revolutionized the way citizens engage with their communities and local governments. Through these innovative tools, individuals can voice their opinions, collaborate on projects, and participate in decision-making processes that affect their lives. This transformation from passive consumption of information to active participation is critical for fostering civic engagement and creating vibrant, responsive communities.

Consider a typical urban neighborhood facing issues like littering or inadequate public amenities. In the past, residents might have expressed their concerns only during infrequent town hall meetings, hoping their voices would be heard. Today, a simple app or website enables those same residents to report problems directly to local officials, share photos, and even propose solutions. This immediate connection not only amplifies their voices but also builds a sense of ownership over local issues.

The power of these digital platforms lies in their accessibility. Whether it's through social media, dedicated apps, or community forums, these tools allow citizens to engage from the comfort of their homes or on the go. For instance, platforms like Nextdoor or Facebook Groups enable residents to connect with their neighbors and share real-time updates about community events, safety alerts, and local initiatives. This instantaneous communication fosters a sense of belonging and collective responsibility, essential for strong community bonds.

Moreover, digital platforms streamline the process of civic engagement by providing user-friendly interfaces that demystify government procedures. Many municipalities have begun to adopt applications that allow constituents to easily access public records, vote on community projects, or participate in public surveys. For instance, platforms like CitizenLab enable local governments to create participatory budgeting processes, where citizens can propose and vote on budget allocations for community projects. This level of involvement encourages transparency and accountability in governance, as officials must consider the preferences of their constituents.

Education also plays a crucial role in enhancing citizen engagement. Many platforms offer resources that inform residents about local governance, civic rights, and the importance of participation. Interactive webinars, informative articles, and video content can equip citizens with the knowledge needed to navigate complex government systems, empowering them to advocate for their needs effectively. Engaged citizens are not only more likely to vote but also to participate in local initiatives, attend community meetings, and engage in discussions about relevant issues.

Youth engagement is another significant aspect made possible by digital platforms. As younger generations are increasingly tech-savvy, leveraging these platforms allows local governments to tap into the energy and creativity of youth. Programs that encourage school students to participate in civic projects via online forums can spark long-lasting interest in community issues. Initiatives like online debates or virtual town halls focus on subjects that resonate with young people, such as environmental sustainability, education reform, or social justice. When young people see their ideas taken seriously, it fosters a sense of agency that can lead to lifelong civic participation.

While the promise of digital engagement is immense, challenges remain. Not everyone has equal access to technology, and digital divides can exclude marginalized groups from the conversation. Therefore, local governments should implement strategies to ensure that engagement opportunities are inclusive. This might involve hosting in-person workshops in community centers or libraries that teach digital literacy. By bridging these gaps, every voice can be part of the dialogue, ensuring that the perspectives of all community members are considered.

Another challenge is the authenticity of engagement. As platforms evolve, so do the tactics employed by individuals or organizations seeking to manipulate public opinion. Misinformation can spread rapidly online, leading to mistrust in civic processes. To combat this, civic tech platforms should prioritize authenticity checks and provide tools for users to verify the information before sharing. Building trust is essential;

if residents do not believe in the integrity of the platform, they will be less likely to engage.

Additionally, continuous feedback loops between citizens and government agencies are vital for sustaining engagement. When the public participates in discussions or shares their concerns, they want to see tangible outcomes. Local governments must close the loop by updating residents on the progress of initiatives and how their input influenced decisions. For example, if citizens proposed a new park, updates on the planning stages, funding allocation, and construction timelines should be communicated effectively. This visibility reinforces the value of citizen contributions and motivates continued involvement.

Community organizers can play a crucial role in fostering engagement through these platforms. By acting as intermediaries, they can help facilitate discussions, organize events, and ensure that diverse voices are amplified. Grassroots organizations can utilize digital tools to mobilize support for community goals, using social media campaigns to raise awareness about local issues or upcoming civic events. Collaboration amplifies impact, creating a synergistic relationship between digital platforms and community organizing.

Cross-sector partnerships can enhance the effectiveness of citizen engagement initiatives. By collaborating with nonprofit organizations, businesses, and educational institutions, local governments can broaden their outreach efforts. For example, a partnership with local universities could result in student-led projects that harness academic resources to address community challenges. These alliances can foster innovation and leverage diverse expertise, enriching the civic engagement landscape.

As technology continues to evolve, emerging tools promise to deepen citizen engagement further. Applications utilizing augmented reality (AR) or virtual reality (VR) can create immersive experiences that help residents visualize proposed projects or changes in their neighborhoods. Imagine walking through a virtual representation of a new community center before it's built, allowing residents to provide feedback based on firsthand experience. Such advancements can foster collaboration and inclusivity in ways that traditional methods cannot.

Regular evaluations of engagement strategies are essential to ensure their effectiveness. Gathering qualitative and quantitative data regarding participation levels, demographic breakdowns, and resident satisfaction can yield insights into what works and what requires improvement. Local governments should adapt their approaches based on this feedback, recognizing that citizen engagement is a dynamic process that may evolve over time.

Finally, celebrating successes reinforces the value of civic engagement. Through newsletters, social media shout-outs, or community events, acknowledging the contributions of engaged citizens fosters a community culture that values participation. Recognizing individual and collective efforts not only motivates further engagement but also helps build lasting relationships between residents and their local governments.

Digital platforms have opened up new avenues for citizen engagement, enabling individuals to take an active role in shaping their communities. By harnessing these tools thoughtfully, local governments can promote transparency, inclusivity, and responsiveness. Ultimately, strong civic engagement strengthens democracy; when citizens feel heard and valued, they become more invested in the health and future of their communities. Through collaboration, education, and continuous feedback, a culture of participation can be cultivated, ensuring that every voice contributes to the collective narrative of society. This engagement not only enhances individual lives but also cultivates a thriving, resilient community that embraces the complexity of modern governance.